Tesla the pride of Electric-Hybrid vehicles:

The unveiling of the Tesla Model 3

Contents

1. Introduction
2. Subjects/Themes
   - **What is Tesla?**
   - **Characteristics/Details of Tesla**

Customer Service

Speed

Charging

Safety

Autopilot hardware

Cost

Time of Sale

Sales

Space

Drive

Roof

Storage

Environmental/Petrol companies

License Plates

Other Tesla Models

Roadster

Model S

Model X

Competition/Motor Vehicles similar to Tesla

Views and Opinions of Tesla customers

Problems with Tesla

   3. **Bibliography**

1. Introduction

Motor vehicles are progressing, improving and developing, with Electric-Hybrid car, the Tesla Model 3, close to being unveiled in 2017.

It was a similar situation with the Toyota's offshoot, the "Lexus LF-LC", being unveiled at the 2012 Sydney Motor Show, and first displayed at the 2012 Detroit Motor Show (Zalstein, David. October 15, 2012).

There are people who want and love cars and motor vehicles, like motor enthusiasts, and people who need cars for necessity and everyday usage. Tesla fits both these groups.

The new Model 3, "could shake up the EV [Electric Vehicle] market worldwide, as well as in Australia." (Tesla Motors). Bullen states, how, "People around the world have been lining up in their thousands." (Bullen, J. April 1, 2016)

Some problems Tesla Motors has to face, is "low petrol prices, high battery costs and uncertain investment in recharging infrastructure." (The Australian, April 1, 2016)

1. Subjects/Themes

What is Tesla?

Tesla is a unique motor vehicle brand that is an electric-hybrid car. Electrical cars are still to make their mark on the motor vehicle market and customers, being a relatively new phenomenon, following on from petrol vehicles. An engineer, in 1888, Nikola Tesla, was the inventor who inspired the company's name. In 2003, a group of engineers in Silicon Valley in America, "wanted to prove that electric cars could be better than gasoline-powered cars, and this is how Tesla Motors was founded." (Tesla Motors)

Tesla Motors states how, "Telsa is not just an automaker [motor vehicle], but also a technology and design company, with a focus on energy innovation. (Tesla Motors) Tesla Motors operates in around 30 countries. The Tesla Model 3 was unveiled at the Tesla factory and design studio in Hawthorne, California.

Golson states how Tesla Motors CEO, and tech guru and billionaire, Elon Musk, highlights how Tesla's new model, Model 3, is "the pinnacle of the Tesla Motors master plan, and, a culmination of a decade's worth of work." (Golson, March 31, 2016/Bullen, J. April 1, 2016) CEO Musk played an important part in producing the hype and investment in Tesla, as a billionaire investor, especially in those times Tesla was struggling (Eisler, M. November 17, 2015). Golson states, how Tesla, "plans to more than double the size of its dealership and service network by the end of next year, in 2017." (March 31, 2016) Tesla Motors wanted to produce a, "mass-market affordable vehicle," in the Tesla Model 3.

Characteristics/Details of Tesla

Customer Service

Tesla has gone over and above, tending to its current and prospective customers. Eisler mentions how, "Tesla's customer service is legendary, and central to the company's construction of brand loyalty." (Eisler, M. November 17, 2015). Eisler goes on to say, how Tesla has involved its customers in the whole experience of launching and promoting and driving its Tesla range. There is Tesla's high cost of delivering and sustaining such a high standard of customer service. (Eisler, M. November 17, 2015).

Tesla Motors is aiming to attract conservative, potential buyers, with 'liberal, romantic, nostalgic lifestyle values', who have very high standards and expectations, with cost effectiveness, comfort and convenience. (Eisler, M. November 15, 2015) Such customers can deal with the possible high cost and initial trouble-shooting of EV's. In stating this, Tesla has to admit there are some problems with their range.

Speed

The Model 3 can travel, at "Zero to 100km/h, under 6 seconds." (Tesla Motors). The Model 3 is slower, and geared for safety, compared to Tesla's previous models: the Roadster, Model S and Model X.

Charging

An electric charge-supercharge for the lithium ion battery of Tesla, can last "345km per charge." Golson states how there are 7,200, high-speed Supercharging networks, in North America, Europe, Asia Pacific, and around the world. The stations provide convenient and free access high speed charging, taking around 20 minutes to charge. (Tesla Motors). Tesla customers can charge at their own home, or one of these 7,200 charge-supercharge stations around the world. (Tesla Motors) Bullen states, how the external supercharger stations around the world, can pump up an EV in minutes, compared to hours taken for an ordinary outlet, like at a home-private residence (Bullen, J. April 1, 2016).

Safety

The Tesla Model 3 has a "5-star safety rating". (Tesla Motors) Model 3 travels slower than the other Tesla models, being geared for safety. Tesla also boasts of having "zero emissions", making it a safe, environmentally vehicle. Tesla Motors CEO, Elon Musk, states how, the Tesla 3, "will be one of the safest cars in the world." (Tesla Motors) Abuelsamid states how, "it's more fun to drive a slow car fast than to drive a fast car slow...I [Abuelsamid], as someone who has driven many different cars, and I can attest to its [Tesla's] accuracy" (Abuelsamid, S. October 16, 2015)

Autopilot hardware

The Tesla Model 3 is computer-rated where it can warn and drive itself. Tesla Motors calls this, "Autopilot Hardware" (Tesla Motors) The electric-hybrid motor vehicle, LeEco, from Beijing, China, has similar autopilot software like the Model 3. (Dunn, M. April 22, 2016)

Cost

Tesla has reduced the cost of its Tesla Model 3, to "US$35,000" compared to other models, like the Tesla Model S and Model X. Tesla is asking for $1,500 to be put down as a pre-order for the Model 3. The Model 3 is much cheaper than its former Tesla models, the Roadster, Model S and Model X, which sale around $100,000. (Golson, J. March 31, 2016) If one thought the Tesla Model S was affordable, the Tesla Model 3, is even more affordable.

However, in a mixed affair, mimajor3, a Telsa enthusiast and buyer, states how, the Model 3 will save buyers in the long term, compared to other similar brands and models. In the short term though, there are less expensive models, than the Model 3, such as the '2015 Hyundai Santa Fe Sport MSRP' is going for $33,000, and the 'Prius Four' MSRP is selling for $28,435. Both are more affordable than the Model 3, which is being offered at $35,000. (mimajor-Telsa, December 19, 2014). Mimajor3 has an "expected term and warranty of around 8 years for each vehicle…but no repairs and maintenance is incorporated into this calculation." (mimajor3, December 19, 2014). Eisler supports mimajor3, by stating how, "Tesla Motors are offering a very generous eight-year unlimited mileage warranty, for battery and driving, for Model S and Model X users." (Esiler, M. November 17, 2015)

Tesla also wants to reduce the cost of its, "lithium ion battery packs" (Tesla Motors). Tesla Motors, how, "by 2020, Tesla's gigafactory, in Nevada, USA, will produce more lithium ion cells than all of the world's combined output in 2013." This will help reduce energy costs for businesses and residences, and provide a back-up power supply. (Tesla Motors) The Australian states how Tesla's massive facility can cut costs of its battery pack by 30%, helping to make an affordable vehicle.

Time of Sale

The Tesla Model 3, will go on sale by the end of 2017, with already, "130,000 pre-orders" around the world (Golson, J. March 31, 2016). Tesla initially launched in December 2014 in Australia. Fans even queued and camped out overnight outside Tesla stores, almost like the Apple Inc, smartphone trend (The Australia, April 1, 2016). Eisler adds to this, stating how Tesla devotees are similar to the Apple Inc. trend, having a strong identification with Tesla's technologies, even them personally promoting Tesla on behalf of the company. However, Eisler is concerned how long and effect will this 'early adopter/buyer enthusiasm' for Tesla last (Eisler, M. November 17, 2015).

Sales

Tesla wants to sell 500,000 cars in general per year by 2020 (The Australian, April 1, 2016). The Australian said how Tesla's "lofty stock/shares price, have jumped in recent days in anticipation of the Model 3 launch." The Australian states Tesla's shares are around $US230, after a year low share price of $US141.05 (April 1, 2016). (The Australian, April 1, 2016) Eisler supports this, stating how Elon Musk is promising a US$35,000, 200-mile Model 3, by the end of 2017-2018, currently increasing the stock valuation of the company. (Eisler, M. November 17, 2015)

The Australian states, how the sale of all-electric and hybrid vehicles was down nine per cent to 60,384 vehicles in 2016 (April 1, 2016). Ford, and its F-series, sold more than Tesla, and other the hybrid and electric vehicles, in February, 2016.

Space

The Tesla Model 3 sits five people comfortably, much like other two-door types of cars. (Golson, J. March 31, 2016).

Drive

The Model 3 comes in "rear-wheel drive and all-wheel drive versions." (Golson, J. March 31, 2016). This, along with the slower speed of the Model 3, allows for more safety and control, compared to the other Tesla models.

Roof

There is a "front to rear roof area, from the windshield all the way to the trunk." (Golston, J. March 31, 2016)

Storage

Like the roof area of the Model 3, the model has more cargo capacity, with front and rear trunks for storage. Apparently, a "7-foot long surfboard" can fit in the Model 3 (Golson, J. March 31, 2016)

Environmental-Climate Change/Petrol companies

Tesla's engineers wanted to see if electric-hybrid motor vehicles, could be better than petrol-oriented motor vehicles. (Tesla Motors) People and groups worldwide, in general, like Tesla, are taking more serious consideration to have its products and services be more environmentally-climate change friendly. This is especially with the pollution from petrol-driven motor vehicles. Tesla Motors CEO, Elon Musk, calls this 'sustainable transportation'. Abuelsamid even goes to the extent of classifying Tesla, as a "green car" (October 16, 2015). Tesla, and its competitors, have been a long time waiting, as electric-hybrid motor vehicles are indirectly in competition with petrol-driven motor vehicles and petrol companies.

It would be interesting, and highly important, to ask and consider the impact, views and intervention of petrol companies on electric-hybrid motor vehicles, like Tesla and its competitors. Tesla might consider involving such petrol companies in their business decisions and actions. Alternatively, petrol companies might create their own unique electric-hybrid motor vehicle, either together with or separate from companies like Tesla Motors.

License Plates

Some decorated Tesla owners have their own unique number plates to match in with the style and culture of various Tesla models.

Other Tesla Models

Tesla Motors has three other models on sale to the global market: Tesla Roadster, Model S and Model X.

Roadster

Tesla Motors states how the Tesla Roadster was brought out in 2008. The Tesla Roadster set a new standard for electric motor vehicles. It can accelerate from 0 to 60 mph in 3.7 seconds, and can achieve a range of 245 miles per charge of its lithium ion battery. Eisler was, "startled by how quiet the car [Roadster] is at cruising speed, even with canvas roof panels removed." People are almost flung back by the fast acceleration of the Roadster (Eisler, M. November 17,

2015) There are now "2,400 Roadsters on the road, in around 30 countries" (Tesla Motors).

In 2008, Abuelsamid was one of the first users of the Roadster, other than Tesla staff. Abuelsamid thoroughly enjoyed the 'performance-responsiveness capabilities' of the Roadster and Model S, comparing it a Lotus Elise. Abuelsamid said there was, "excellent suspension, big tires and a low center of gravity...immense instant torque [especially Model S], with lithium ion cells helping to make better Tesla vehicles" (October 16, 2015). However, Abuelsamid said, after a while, driving both Tesla Roadster and Model S, is not something he would do often. This is similar to contented Tesla owner-user, Darrell, who preferred his Model S on weekdays instead of his Roadster, which he used on weekends.

Model S

Tesla Model S appeared in 2012 and was one of the world's top quality electric sedans, in fact, 100% electric. Can one believe, the Model S can fit seven people, as a four-door car. It is one of the fastest four-doors ever made, with the acceleration of a sports car, travelling from 0 to 60 mph in 3.2 to about 5 seconds. Model S has a high efficiency motor. It is also a family sedan. It has a "low center of gravity, allowing for outstanding road holding and handling." (Tesla Motors) Model S achieves "265 miles per charge". It was named 'Motor Trend's 2013 Car of the Year', and given a '5-star safety rating' from the U.S. National Highway Traffic Safety Administration. (Telsa Motors). Eisler extends these awards of the Model S, with 2014 Consumer Reports rating the Model S, 'the best in the show'. (November17, 2015). Model S has more than 50,000 on the road worldwide.

The Model S has "two dual motor all-wheel drive", with, both rear-wheel drive and all-wheel drive versions, like the Model 3. Model S has a 17-inch portrait touchscreen, whereas the Model 3 has a 15-inch touchscreen. Like the Model 3, Model S has front and rear trunks of 64 cubic feet of storage. Model S, like Model X, is much more expensive than the Tesla 3, at around $100,000. Model S and Model X buyers helped contribute funding to the development of the Model 3. (Golson, J. March 31, 2016). Eisler supports this, "with the sale of premium EVs funding development of a battery electric vehicle for the everyday person." (Eisler, M. November 17, 2015)

Model X

Model X is similar to a SUV. Tesla Motors describes Model X as "the safest, fastest and most capable sport utility vehicle in history." It is an all-wheel drive; has a 90 kWh battery, providing 470 km or range; comfortably seats seven people, similar to the Model S; and accelerates from zero to 100 km per-hour in as quick as 3.4 seconds, which is very fast (Tesla Motors).

Tesla Motors states, how Model X is a crossover vehicle. Model X has "exhilarating acceleration, falcon wing doors (with doors similar to attractive sports cars, like the Lamborghini) and three-row seating." (Tesla Motors)

Competition/Motor Vehicles similar to Tesla

The Australian states, how, "Model 3 has entered a crowded field of luxury and electric cars." (April 1, 2016). There are similar to EV (Electric Vehicles) on the market from other brands: the Nissan Leaf, LeEco, Faraday Future, Audi A4, BMW 3-Series, Chevrolet Bolt EV and Volkswagen e-Golf. Tesla Motors is very similar to LeEco from Beijing, China, as also being a design and innovation company.

Golson states, how General Motors (GM) 'Chevrolet/Chevy Bolt', is "the biggest direct competitor to the Tesla Model 3." The Chevy Bolt has a similar price range to the Model 3, at around $30,000, and has an electric charge range of 200 miles per charge (Golson, J. March 31, 2016). The 'Nissan Leaf' claims to offer more driving range at a similar price to the Model 3 and Chevy Bolt. Eisler supports how Nissan has provided strong performing battery cell technology, despite sales of its Leaf EV dropping. (Esiler, M. November 17, 2015)

The 'LeEco' concept car, or 'Le Super Electric Ecosystem', from Beijing, China, is another main competitor to the Tesla Model 3. The LeEco, "has smartphone voice recognition software (automated driving and parking), on-board entertainment system...a speed of 210km/h, a foldaway steering wheel, exterior display on the front dash of the car...back of seat displays, isolated music playback." (Dunn, M. April 22, 2016) The company behind the LeEco, also deal in smartphones, TV set-top boxes and smart TVs. LeEco is also supporting an American-version electric car start-up, the 'Faraday Future'. Additionally, LeEco is partnering with Aston Martin, to bring out similar electric car technology and infotainment. (Dunn, M. April 22, 2016).

Abuelsamid states, that although he loves the Tesla Model S, if he had the budget, he would select, the 'Volkswagen e-Golf' (October 16, 2015).

Views and Opinions of Tesla customers

In general, Tesla, and similar vehicles, are liked by most customers and critics, with some liking and some having different opinions of Tesla and its range, and EVs. Under Eisler, 'Darrell' has two Telsa vehicles he is passionate about, with the Roadster on weekends and weekdays, the Model S. Darrell is a technology-buff, working in IT administration, having that connection with the Tesla Model ranges. Darrell is one of many Tesla drivers, with "a sense of adventure…and putting themselves on the line to try new technologies, as with Tesla." (Eisler, M. November 17, 2015) Tesla's customers, like Darrell, are helping to shape Tesla, motor vehicle and technological history. This is especially with Tesla most welcomely and intimately including its customers in its whole sales and marketing showcasing and experience.

The majority of vehicle enthusiasts, including Tesla, would not considering speeding and thrashing their Tesla or non-Tesla at the local drag strip. Although, this would be a temptation, as the acceleration of Tesla models is there for the awaiting-experience, like the Tesla Roadster for instance. Some Tesla owners and users might go down country roads, or take the long, joy-ride home, instead of the drag strip, which will not land the user in jail (Abuelsamid, S. October 16, 2015)

Problems with Tesla

After the strong and positive 2015 Consumer Report of the Tesla Model S, two months later, a number of problems arose for Model S users: "poor fit and finish, leaking battery cooling pumps and warped brake rotors that would be costly to fix out of warranty." (Eisler, M. November 17, 2015). First time users of Tesla and EVs, have the trouble of journeying an unknown, good-problematic track of EVs.

Battery warring and ageing, is the main issue and problem for Tesla. Eisler states how batteries degrade over time, possibly explain the large eight-year warranty. Tesla, and other brands, need to regulate the 'parts supplier', which look to earn the most from such batteries, either becoming suppliers themselves, or organizing a good deal with other relevant suppliers.

(Eisler, M. November 17, 2015). Hybrid cars "require smaller and less costly powerpacks compared to battery-only EVs." (Eisler, M. November 17, 2015). A mixture of Hybrid-Electric vehicles is a better match, as such vehicles, can last for decades.

But the 'pure battery electric cars', have hidden replacement costs that consumers may or may not want to shoulder. This means, as stated before, targeting potential customers who do expect high standards and can endure the high costs from time-to-time and initial problems of early EV technology. (Eisler, M. November 17, 2015).

Mimajor3 stated how Tesla has less expenses and costs in the long term, with Hyundai and Prius being more affordable short term. (mmajor3, December 14, 2014)

2. Bibliography

- Abuelsamid, Sam. (October 16, 2015) 'Car buffs don't all dislike Tesla – They just have different priorities,' <u>Forbes</u>
- Bullen, James. (April 1, 2016) 'Tesla unveiling new Model 3 electric vehicle in California', <u>HuffPost Australia</u>
- Dunn, Matthew. (April 22, 2016) 'Chinese tech giant has unveiled a new electric car it hopes will rival the Tesla Motors', <u>News.com.au</u>
- Eisler, Matthew. (November 17, 2015) 'Can Tesla's enthusiast customers help it sell the electric car for the everyperson?' <u>The Conversation</u>
- Golson, Jordan (March 31, 2016) 'Tesla Model 3 announced: release set for 2017, price starts at $35,000', <u>The Verge</u>
- Mimajor3 (December 19, 2014) 'Tesla Model 3 Enthusiast and Buyer', <u>Tesla Motors: Forums</u>
- Tesla Motors Australia, 'About Tesla'
- Tesla Motors Australia, 'Model 3'
- Tesla Motors Australia, 'Model X'

- The Australian (April 1, 2016) 'Tesla Model 3 unveiled: an affordable electric car for the masses?'
- Zalstein, David. (October 15, 2012) '2012 Sydney Motor Show Preview,' CarAdvice.com.au